Making Transatlantic Defense Cooperation Work

Findings and Recommendations of the
CSIS *Atlantic Partnership* Project

Making Transatlantic Defense Cooperation Work

Findings and Recommendations of the CSIS *Atlantic Partnership* Project

Steering Committee Cochairs

The Rt. Hon. Sir Geoffrey Pattie, MP

Michael T. Smith

Research Staff

Daniel Gouré

Robin Niblett

David C. Earnest

CSIS

The Center for Strategic & International Studies Washington, D.C.

About CSIS

The Center for Strategic and International Studies (CSIS), established in 1962, is a private, tax-exempt institution focusing on international public policy issues. Its research is nonpartisan and nonproprietary.

CSIS is dedicated to policy analysis and impact. It seeks to inform and shape selected policy decisions in government and the private sector to meet the increasingly complex and difficult global challenges that leaders will confront in the next century. It achieves this mission in three ways: by generating strategic analysis that is anticipatory and interdisciplinary, by convening policymakers and other influential parties to assess key issues, and by building structures for policy action.

CSIS does not take specific public policy positions. Accordingly, all views, positions, and conclusions expressed in this publication should be understood to be solely those of the authors.

Library of Congress Cataloging-in-Publication Data

CSIS Atlantic Partnership Project.
Making transatlantic defense cooperation work : findings and recommendations of the CSIS Atlantic Partnership Project / steering committee cochairs, Geoffrey Pattie, Michael T. Smith : research staff. Daniel I. Gouré, Robin Niblett, David C. Earnest.
p. cm. — (Panel reports, ISSN 0899-0352)
Includes bibliographical references (p.).
ISBN 0-89206-286-X
1. Defense industries—Europe. 2. Defense industries—United States. 3. Defense industries—International cooperation. 4. Technology transfer—International cooperation. 5. Military research—International cooperation. 6. CSIS Atlantic Partnership Project. I. Pattie, Geoffrey. II. Smith, Michael T., 1943– III. Center for Strategic and International Studies (Washington, D.C.) IV. Title. V. Series: CSIS panel reports.
HD9743.E922C68 1996
327.1'7—dc20 96-14000
CIP

ISBN 0-89206-286-X

The Center for Strategic and International Studies
1800 K Street, N.W., Washington, D.C. 20006
Telephone: (202) 887-0200
Fax: (202) 775-3199
E-mail:books@csis.org
Web site: http://www.csis.org/

Contents

Glossary

Acronym	Definition
ACTD	Advanced concept technology demonstrator
AEW	Airborne early warning
AFV	Armored fighting vehicle
AGS	Airborne ground surveillance
AI	Artificial intelligence
AMRAAM	Advanced Medium-Range Air-to-Air Missile
APGM	Advanced Precision Guided Munitions program
ASRAAM	Advanced Short-Range Air-to-Air Missile
ASTOR	Airborne Stand-Off Radar System (U.K. airborne ground surveillance)
AWACS	Airborne Warning and Control System
BSM	Battle space management
BM	Battle management
BM/C^3	Battle management/command, control, and communication
BPI	Boost-phase intercept
C^3	Command, control, and communication
C^4I	Command, control, communication, computers, and intelligence
CNAD	Conference of National Armaments Directors
CONUS	Continental United States
Corps SAM	Corps surface-to-air missile
Creso	Italian heliborne systems (airborne ground surveillance)
DOD	Department of Defense

Acronym	**Definition**
ACTD	Advanced concept technology demonstrator
DFARS	Defense federal acquisition regulation supplements
DSP	Defense satellite program
EAD	Extended air defense
ECCM	Electronic counter-countermeasures
ESSM	NATO Evolved Sea Sparrow Missile
EU	European Union
FLA	Future Large Aircraft
IFF	NATO Identification-Friend-or-Foe system
JAST	Joint Advanced Strike Technology
JTIDS	Joint Tactical Distribution System
JSTARS	U.S. Joint Surveillance and Target Attack Radar System
MBT	main battle tank
MEADS	Medium Extended Air Defense System
MIDS	Multifunctional Information Distribution System
MOU	Memorandum of Understanding
MSOW	Modular Stand-Off Weapon program
NAGS	NATO airborne ground surveillance
NATO	North Atlantic Treaty Organization
NBC	nuclear, biological, and chemical
R & D	Research and development
RFP	Request for proposal
THAAD	Theater High-Altitude Air Defense System
TMD	Theater missile defense
WEU	Western European Union

Executive Summary

Why Have We Undertaken the CSIS Atlantic Partnership Project?

As defense industry leaders, we approach the subject of transatlantic defense cooperation with great caution. There have been too many past failures for us to do otherwise. Nevertheless, we have undertaken this project because we believe that we now stand at the threshold of a markedly changed era of transatlantic defense-industrial relations. There are two central challenges that we must confront.

First, the continuing and severe cuts in defense budgets on both sides of the Atlantic, combined with the spiralling cost of developing many new defense technologies, mean that the United States and Europe will find it prohibitively expensive to satisfy all their military requirements on a "go-it-alone" basis. With defense spending unlikely to increase, U.S. and European governments must get the most out of their defense investments. Transatlantic collaboration on programs can spread the risks of technology development and provide economies of scale in production and logistics. Collaboration will work, however, only if we can find new incentives and give these programs a better chance of success within the political and legislative process.

There is a second challenge. In the next few years, U.S. and European governments and industry will either develop a new, more reciprocally beneficial relationship between their defense industrial bases, or the present course of defense industrial consolidation could lead to the emergence of separate and adversarial defense markets on either side of the Atlantic. We must avoid polarization of U.S. and European defense markets if industry is to meet the military and technological requirements of its governments at acceptable cost levels.

The CSIS Atlantic Partnership project has developed two sets of recommendations from industry's perspective to deal with these challenges. We recognize that these recommendations fall within the context of an ongoing transatlantic dialogue on defense cooperation. We believe, however, that these recommendations offer practical steps forward because they recognize the structural nature of some of the obstacles to transatlantic defense cooperation. Instead of developing a wish list of political changes, this project has concentrated on ways to strengthen the incentives for cooperation.

Structural Obstacles to U.S.-European Defense Cooperation

- Instinctive preferences for self-sufficiency

- Difficulty of agreeing upon common requirements
- Conflicting arms export policies
- Multiyear (European) versus annual (U.S.) approaches to funding
- Imbalances between U.S. and European defense markets and budgets
- No European interlocutor for collaborative programs

Recommendations

A Framework Agreement for Transatlantic Collaborative Programs

U.S. and European governments have reaffirmed their interest in pursuing certain defense programs on a collaborative basis, such as from the development phase onward. If such programs are to succeed, they must offer incentives and allow industry to offer the best solutions in terms of technology development, production, and program management. The Task Force recommends that U.S. and European governments immediately develop—and their legislatures approve—a Framework Agreement for collaborative programs, which would overcome several of the principal obstacles to transatlantic defense cooperation.

Governments with collaborative programs under the Framework Agreement, would agree to the following five action items:

- Coordinate their rules and procedures for technology transfer and security, so U.S. and European companies can share technology under normal industry-contractual and proprietary rules.
- Allow participating companies to determine the distribution of work content, technology development, and program management, including arrangements for prime contractorship in accordance with the parameters for national investment and return that governments have negotiated up front.
- Assure stability and continuity for the cooperative programs through clearly stated funding milestones, including the financial liabilities for governments that end their participation.
- Agree to an accounting system for contributions to the program, including an assessment of the impact of currency adjustments.
- Enable participating companies to export the collaborative system without further legislative restrictions to those third-country destinations that governments have identified and agreed upon in advance.

A collaborative program would first have to satisfy several prerequisites to be eligible to proceed under the Framework Agreement. U.S. and European governments must be able to

- identify a program for collaboration. Such a program would
    - meet a carefully coordinated set of common military requirements;
    - be based upon complementary technological capabilities and not duplicate research and development (R&D), production, and other costs; and
    - improve the standardization and interoperability of the forces of participating governments.
- decide upon the levels of national investment in and return from the program, including the selection process for awarding the contract.
- agree on the arrangements for exporting the final system. This would involve
    - establishing a common list of destination countries to which industry would, in principle, receive approval to export, and
    - accepting that modification of that list would require the unanimous approval of participating governments.
- secure approval from their legislatures to proceed with the collaborative program under the terms of the Framework Agreement.

The Task Force recognizes that only a few programs could proceed under the Framework Agreement and supports programs in the areas of missile defenses, communications and identification systems, land combat systems, sea lift and sea propulsion, and battle-space management (see Annex A).

A Strategy of Constructive Engagement between U.S. and European Defense Markets

Historically, the unbalanced relationship between U.S. and European defense markets (in terms of budget size and national markets, as well as procurement styles and policies), has made it extremely difficult to open the transatlantic market in the defense field. Currently however, governments and industry in the European Union (EU) are moving from a culture of fragmentation to one of greater integration in the defense industrial sector. This transition offers some important opportunities for transatlantic defense cooperation:

- More effective European coordination of national military requirements will allow European governments to be effective interlocutors with the United States on collaborative programs.
- With greater consolidation, European companies often will be in a better position to act as genuine partners with U.S. companies on cooperative programs.

- Greater consistency in national European procurement practices and policies will allow companies from both Europe and the United States to operate in a more transparent and predictable market environment.
- If European governments are able to coordinate and formalize their domestic source preferences and procurement policies, Europe and the United States may be able to negotiate opening their markets.

Given these opportunities, U.S. government and industry have good reasons to hold discussions with their allies about the effect of European defense industrial integration on transatlantic defense cooperation.

Constructive engagement will also help lessen the risks inherent in this transition away from opaque national procurement policies and practices and toward a more coherent European defense identity. The growth of interlocking European, cross-border industrial interests, combined with deep-rooted perceptions in Europe of an uneven playing field in transatlantic defense trade, means that existing patterns of U.S.-European defense cooperation, based on large sales of U.S. defense equipment to Europe and on U.S.-led coproduction deals, are not sustainable in the long term. Without an effective U.S.-European dialogue on defense cooperation, therefore, U.S. defense manufacturers may see their traditionally high degree of access to European defense markets diminish as European governments come under increasing pressure to strengthen their domestic defense industries to be competitive with U.S. firms.

To avoid the risks of a new polarization of U.S. and European defense markets, governments on both sides of the Atlantic could undertake some important, reciprocal steps:

In Europe

As European governments pursue the development of a more integrated defense identity, they should ensure that their defense markets remain open to competition and cooperation with the United States. To this end, they should

- avoid instituting new forms of domestic protectionism;
- cooperate to develop a more coherent and transparent European defense market, as is currently being discussed in the Western European Union (WEU), providing that any new common external European barriers do not constitute an overall rise in the obstacles facing U.S. defense imports; and
- continue the process of rationalizing and privatizing European defense companies.

In the United States

The U.S. government should make clear its interest in a competitive defense market that is more open to the companies of allied governments. To this end, it should

- reject attempts to introduce new legislation that would insist on sole-sourcing from U.S. companies in specific sectors of the defense market.
- revise U.S. critical items lists and other mobilization base restrictions to allow European systems to compete in a broader range of U.S. program requirements, especially in those areas in which European companies offer technological advantages and in which the United States does not have the technological capabilities or assets.
- strengthen and extend the exemptions from "Buy America" legislation for allied countries.

Collectively

The existence of formal and informal offset requirements for U.S. companies bidding on European programs is tied to U.S. restrictions on the ability of European companies to bid for U.S. programs as prime contractors. Only a reciprocal effort by governments and industry on both sides of the Atlantic will ease these obstacles to the emergence of a more efficient and competitive transatlantic defense market.

CHAPTER 1

Introduction

As defense industry leaders, we approach the subject of transatlantic defense cooperation with great caution. There have been too many failures for us to do otherwise. Nevertheless, we have undertaken this CSIS project because we believe that we now stand at the threshold of a markedly changed era of transatlantic defense industrial relations. There are two central challenges we must confront.

First, the continuing and severe contraction in defense budgets on both sides of the Atlantic, combined with the spiralling cost of developing many new defense technologies, means that U.S. and European governments will find it prohibitively expensive to satisfy all their military requirements on a "go-it-alone" basis. Development and production of defense systems on a sole national (U.S.) or regional (European) basis will leave governments with the choice of whether to reduce the variety or the capability of the systems they acquire. They will also often pay higher unit costs than necessary for those that they do procure. With defense spending unlikely to increase, U.S. and European governments must get the most out of their defense investments; collaboration on cooperative programs can spread the risks of technology development and provide economies of scale in production and logistics. Transatlantic collaboration will work, however, only if defense industry leaders can find new incentives that give these programs a better chance of success within the political and legislative process.

Figure 1: What Do We Mean by Collaboration?

"Transatlantic defense cooperation" is a notoriously vague term. It tends to refer to the full spectrum of interactions between U.S. and European governments and industries that enable non-domestic contributions to national defense programs. In other words, defense cooperation ranges from the transatlantic trade in defense components and systems, through licensed coproduction, to the collaborative development and production of defense platforms and technologies.

This report uses the term "collaboration" to refer to joint efforts by U.S. and European governments to evaluate, research, develop, and produce a defense system based upon a set of common requirements. It uses "cooperation" as the generic term to cover all other forms of cooperative activities, including defense trade and coproduction.

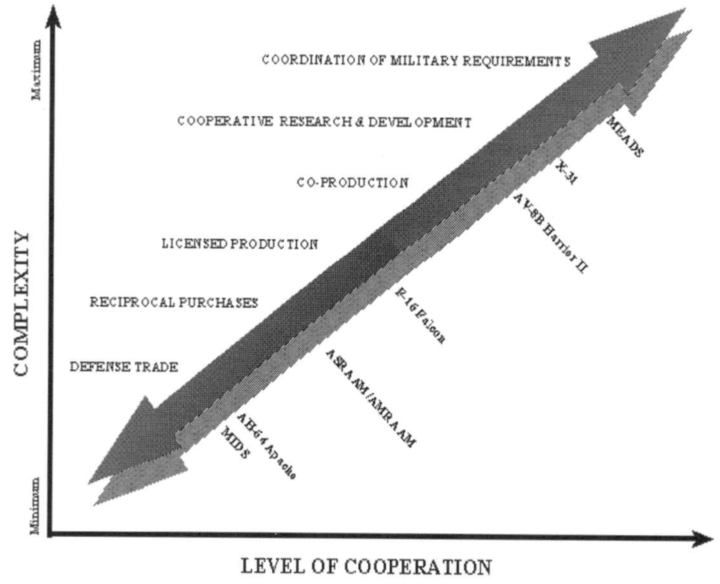

The second challenge for transatlantic defense cooperation is that in the next few years, either U.S. and European governments and industry will develop a new, more reciprocally beneficial relationship between their defense industrial bases, or the current pressures for defense industrial consolidation in the United States and in Europe will lead to separate and adversarial defense markets on either side of the Atlantic. Recent U.S. defense industrial consolidation has created a few, very large, defense conglomerates that will dominate domestic markets and drive competition in global export markets, whether in Europe, the Middle East, or Asia, to new levels of intensity (see figure 2). At the same time, governments and industry in the EU are moving from a culture of fragmentation to one of integration. A mixture of budgetary, political, and operational pressures have turned intra-European defense cooperation into a central focus for European governments. They are looking beyond industrial consolidation at the national level and are encouraging defense consolidation at the European level, including the creation of pan-European consortia and some shelving of national capabilities.

Figure 2: Average Annual Sales of the Top 10 U.S. and Top 10 European Defense Companies, 1989-1993

Despite defense spending reductions since the end of the cold war, U.S. defense companies have remained relatively larger than their European counterparts.

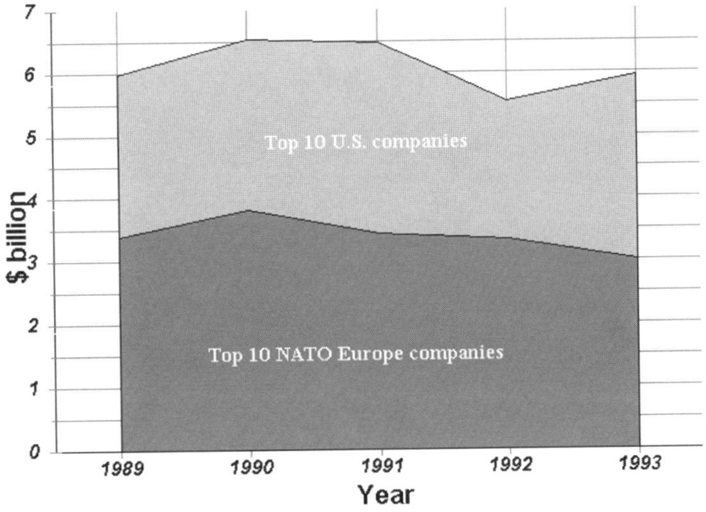

Sources: *SIPRI Yearbooks 1990–1994: World Armaments and Disarmament* (New York: Oxford University Press, 1990–1994).

In the long term, this process of European defense integration will make unsustainable the patterns of defense cooperation now based on large sales of U.S. defense equipment to European governments and on coproduction deals backed by offsets. Although U.S. companies often have been able to work around divergent national procurement policies and practices in Europe, they will find it far harder if the opacity of such practices at the national level is reinforced by the emergence of a "European preference," based on interlocking cross-border industrial interests and on deep-rooted perceptions in Europe of an uneven playing field in transatlantic defense trade. Attempts to stick to the status quo will mean that U.S. defense manufacturers will see their traditionally high degree of access to European markets diminish in the face of European governments' efforts to strengthen their domestic defense industries to be competitive with U.S. firms.

Whatever the potential short-term benefits to certain interests of protectionist and separate U.S. and European markets, we believe that these are far outweighed by the long-term costs that such an environment would impose on industry's ability to meet the financial and military requirements of U.S. and European governments.

The Rationale for Cooperation

The arguments in favor of transatlantic defense cooperation are well rehearsed but are nonetheless compelling.[1] It is worth reviewing them here, particularly given the important shift in the rationale that has taken place since the end of the cold war. In the post-cold war strategic and defense economic environment, the desire to support transatlantic political cohesion is no longer reason enough to justify cooperative initiatives. U.S.-European defense industry cooperation must be based upon long-term strategic interests, mutual economic advantage, and technological synergy. It must also recognize some new realities.

- *U.S.-European military interoperability is as important as ever.* The increasing joint participation of U.S. and European forces in multinational peacekeeping and peacemaking operations puts a renewed premium on standardization of some of their equipment and of protocols at the highest possible technology levels. This standardization can be assisted, in part, through the development, production, and purchase of common equipment. Resolving the interface requirements at the design level, and not on the battlefield, will greatly assist the interoperability of allied forces and the efficiency of logistical support for all sorts of coalition operations.

- *Sustaining our technology edge remains a high priority.* U.S. and European governments place as high a priority as they did during the cold war on maintaining a "technology edge" over any potential adversaries. Maintaining this technology edge is a priority, given threats to U.S. and European security interests in regions as far apart as the Balkans, the Middle East, and East Asia; given the rapid global expansion of military capabilities; and given the decreased Western public tolerance for suffering casualties in conflicts that do not involve the defense of national territory. If cooperation between U.S. and European companies can produce systems of a high technological caliber, then such cooperation should be encouraged.

- *No one can be the best in all technologies at all levels.* While the need for a technology edge is as pressing as ever, U.S. hegemony in high technology across the board is a relic of the past. As a 1988 U.S. Defense Science Board study concluded, "We are, and will remain, dependent on foreign sources for critical components of our weapons systems. We cannot eliminate foreign dependency in this era of a globalized defense industry."[2] Moreover, shielding U.S. high technology from foreign competition and cooperation can obstruct innovation and increase the price of U.S. defense

---

1. See David Abshire, Richard R. Burt, Philip Merrill, William H. Taft IV, and R. James Woolsey, *Transatlantic Defense Cooperation in a Time of Transition* (Washington, D.C.: Center for Strategic and International Studies, December 1992).

equipment. A number of recent studies have found that European companies are capable of making major technological contributions in a number of U.S.-designated critical technologies.[3] By encouraging industry collaboration in areas as diverse as artificial intelligence, water jet propulsion, and composite materials, U.S. and European military forces can gain access to the most sophisticated systems.

❒ *Complete autarky in the development and production of defense systems is no longer feasible*. The governments of the United States and Europe will find it prohibitively expensive to satisfy all critical military requirements on a go-it-alone basis for two reasons: the spiralling cost of developing new defense-related technologies and the simultaneous decline in defense budgets.[4] U.S. acquisition reform, accessing commercial technologies, and the open commitment of the Clinton administration to U.S.-European defense cooperation are all signs that the U.S. government recognizes this fact. It is equally doubtful that European countries can build a cost-effective and technologically dynamic pan-European defense industrial base without some sort of access to U.S. markets, technologies, and finance, as well as exposure to competition and cooperation with U.S. companies. Development and production of defense systems on a solely national (U.S.) or regional (European) basis will mean that governments will have to reduce the variety of systems that they can produce and pay higher unit costs for those that they do procure.

❒ *Cooperation can save money*. At the component level, domestic sole-sourcing fosters an environment for uncompetitive bids. At the program level, the increasing sophistication of defense equipment requires huge up-front investments in new programs. Collaborative efforts by governments through cooperative programs spread the risks of technology development and provide economies of scale in production and logistics. True, the total program costs are often much greater because of the complexities of managing an international program. Yet, each nation's share of that total cost is less than the cost of pursuing a national program. The Anglo-American

---

2. Office of the Under Secretary for Defense Acquisition, *Final Report of the Defense Science Board Summer Study on the Defense Industrial and Technology Base* (Washington, D.C.: U.S. Department of Defense, October 1988), 1.

3. See, for instance, Office of Technology Assessment, *Arming Our Allies: Cooperation and Competition in Defense Technology* (Washington, D.C.: U.S. GPO, May 1990), which lists the significant European capabilities in machine intelligence and robotics, simulation and modeling, weapon system environment, air-breathing propulsion, high-energy density materials, biotechnology materials and processes, and composite materials.

4. U.S. military hardware is depreciating at around $101 billion per year; new investment, however, is running at only $79 billion ($43.3 billion in FY95 dollars for procurement budget authority and $36.2 billion for R&D budget authority). Stephen A. Cambone, Daniel Gouré, and Don M. Snider, *Defense in the Late 1990s: Avoiding the Train Wreck* (Washington, D.C.: The Center for Strategic and International Studies, 1995).

AV-8B program, for example, is estimated to have cost 25 percent more than it would have as a single-nation program. This is still 75 percent less than the cost of two comparable single-nation programs—a saving of over 30 percent for each participant.[5] (See Annex B).

- *Cooperation can also sustain jobs and expertise*. The industrial rationalization that accompanies reduced defense spending in both the United States and Europe inevitably leads to closures of R&D and production facilities and the loss of jobs. The sharing of costs implicit in collaborative programs can offer governments the opportunity to proceed with the development and production of systems that might otherwise be unaffordable on a go-it-alone basis. If such programs are conducted under terms that allow companies to negotiate a fair return in rough proportion to the national investment, then cooperative programs can help governments to maintain employment and technology expertise in areas that might otherwise disappear.

U.S. and European governments and industry have good reasons to want to seize the opportunities inherent in transatlantic defense cooperation. Uniquely, the transatlantic dimension offers governments and industry in the Atlantic Alliance the opportunity to encourage cooperation based upon increasing areas of technological parity, a record of successful industry partnerships at the subcontractor and licensed production levels, and a mutual security treaty that can overcome fears of interdependency.

And yet, the history of defense cooperation is not a positive one. At the codevelopment level, there have been a number of notable failures in transatlantic collaboration.[6] For the reasons described above, members of the Task Force are unanimous in their desire to break with this pattern of failure and pursue practical opportunities to collaborate on specific technologies and programs in response to their governments' requirements. Looking at defense cooperation more broadly, the Task Force recognizes that imbalances in current patterns of transatlantic defense cooperation may be politically unsustainable in the long-term.

Before embarking upon our recommendations to overcome these patterns of failure in codevelopment, trade imbalance, and coproduction, it is necessary, first, to have a good understanding of the obstacles to transatlantic defense cooperation.

Structural Obstacles to Cooperation

It is easy to develop lengthy lists of the many obstacles to U.S.-European defense cooperation, as a precursor to developing recommendations for an improved coop-

5. In 1992, the Department of Defense Inspector General estimated that the United States could save as much as $10 billion over the next five years by cooperating with its allies in developing new weapons systems. See, U.S. Department of Defense, Office of the Inspector General, *International Cooperative Research and Development* (Washington, D.C.: U.S. Department of Defense, October 1992).

6. For example, programs such as NATO Identification-Friend-or-Foe, the Advanced Precision Guided Munition, and the Modular Stand-Off Weapon.

erative environment. Such lists usually refer to, among others, disjointed military requirements, technology transfer restrictions both within and outside the Alliance, "Buy American" laws and European favoritism, a lack of market transparency, and a lack of long-term commitment by governments to cooperative programs.

It is pointless, however, to base recommendations on calls for the wholesale reform or abolition of these obstacles (this has certainly not worked in the past). These obstacles are, in part, symptoms of the different cultures within which U.S. and European governments and industry must operate. These obstacles, however, are also symptomatic of deeper, structural impediments. Effective recommendations will depend upon our ability to discern obstacles to cooperation that are structural from those that are symptomatic. Recommendations must be based on working around the structural obstacles rather than on the assumption that it will be possible to overturn them right away (see figure 3).

Figure 3: Industrial Culture as a Barrier to Collaboration

The different U.S.-European defense industrial cultures can lead to damaging misunderstandings between governments and companies that want to pursue transatlantic collaborative programs.

United States	Europe
• A single market	• Fragmented but integrating
• Domestic market still considered sufficient to sustain independent industry	• Domestic markets insufficient to sustain national industry
• Competitive procurement among U.S. companies in most areas	• A more directed and competitive procurement mix
• Two or more competitors in most technology sectors	• National champions in many sectors and, often, one European industrial alliance
• Arms-length government-to-industry relationship	• National treatment varies widely
• Exports increasingly important to defense industry	• Exports essential
• Government-to-government collaboration the exception	• Transnational collaboration the norm throughout the life of the program

Source: Sydney Gillibrand, Gerald A. Johnston, and Robin Niblett, *The Atlantic Partnership: An Industrial Perspective on Transatlantic Defense Cooperation. Final Report of the CSIS Atlantic Partnership Steering Committee* (Washington, D.C.: The Center for Strategic and International Studies, 1991), 10-11.

Obstacles to Collaborative Programs

The Difficulty of Agreeing Upon Common Requirements. Successful collaborative programs depend upon the ability of governments to coordinate their military requirements as well as on the schedules under which they will replace existing military hardware. Unfortunately, reaching agreement is not simply a matter of efficient procedure, whether within the North Atlantic Treaty Organization

(NATO), in a narrower multilateral context, or even among the U.S. armed services. The end of the military threat in Europe from the Warsaw Pact countries has accentuated the fact that U.S. and European governments often have fundamentally different security priorities, leading to different national military requirements. These differences make it very difficult to come to agreement on the necessary common requirements for collaborative programs.

Conflicting Arms Export Policies. The United States and many European governments use arms exports to support foreign policy objectives and to sustain comprehensive national defense industrial bases. Although U.S. and European governments share a concern about the proliferation of conventional weapons, they have often disagreed about which destination countries constitute threats to global or regional security. In the past, for example, European defense-related products have been exported to nations toward which the United States has maintained national embargoes. Until the U.S. and European governments can better coordinate their arms export policies, there is no prospect for a permanent, across-the-board liberalization of national export control legislation to facilitate greater defense cooperation and overcome the issue of "extraterritoriality."

An Absence of Stability and Continuity (United States). Whereas European governments often make multiyear commitments to defense programs, the U.S. legislative process allows the Congress to review annually all weapons programs. Collaborative programs, which are perceived to offer a smaller "national" return and which generally feature more complex and costly management oversight structures, are often the first to fail the scrutiny of congressional budget-cutters. The unpredictability of the annual U.S. budgetary review process acts as a considerable disincentive to companies that are considering whether to enter into a collaborative program.

An Absence of Stability and Continuity (Europe). There is no cohesive European "actor" that can serve as an interlocutor with the United States on collaborative programs. Although European governments have made considerable progress in identifying common military requirements and in making cross-border programs more manageable, challenges defining common specifications, managing collaborative programs, and synchronizing the procurement time lines of different countries still exist.

The Imbalance Between U.S. and European Defense Budgets. The fact that U.S. defense contractors have access to national defense procurement and R&D budgets that outstrip the combined budgets of NATO's European members can act as a structural disincentive to collaborative programs (see figure 4). U.S. companies have rarely needed to mitigate the costs and risks of developing defense systems through international cooperation. In the case in which transatlantic collaboration may be an option from a cost and requirements perspective, the frequent discrepancy between the size of U.S. and European procurement requirements can make it very hard to build balanced collaborative programs that offer "fair" returns to participating governments and companies in terms of technology development.

Figure 4: United States versus NATO Europe Total Defense Expenditures, 1985–1994

U.S. defense industries operate from a foundation of larger domestic defense budgets than all their NATO Europe counterparts combined. U.S. defense expenditures have been approximately twice the aggregate total defense expenditures of four of Europe's largest arms-producing countries: France, Germany, Italy, and the United Kingdom.

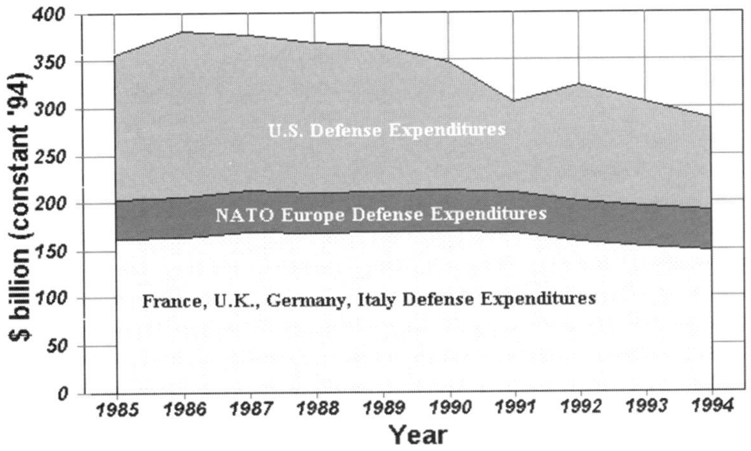

Source: *SIPRI Yearbook 1995: Armaments, Disarmament and International Security* (New York: Oxford University Press, 1995).

Obstacles to Defense Trade

The Preference for Self-Sufficiency. Both U.S. and European governments have an instinctive preference to go it alone in defense production for a host of reasons. These reasons include the desire not to have to compromise on national military requirements and the need to limit dependence on foreign suppliers, support national high technology research, and channel taxpayers' money back into the national economy. Such objectives create a bias in favor of domestic sourcing. As a result, the United States still has explicit regulations and statutes requiring that federal government procurements be made from domestic manufacturers. In Europe, the need to transcend fragmented national defense markets and, at the same time, the desire to sustain national defense industries, are creating pressures for a "buy European" preference and intra-European reciprocal purchases.

Today, while budgetary pressures have sharpened the need for defense cooperation, the desire of governments to retain as much self-sufficiency as possible has not diminished. This will affect the future pace and content of initiatives toward true reciprocal opening of defense markets.

The Imbalance Between U.S. and European Defense Markets. The mismatch between the unitary nature of the U.S. defense market and the fragmented nature of European defense markets constitutes the central structural obstacle to reciprocal market-opening initiatives on either side of the Atlantic. In Europe, U.S. companies must cope with an array of differing and opaque national defense procurement polices and practices. The less formal and legalistic approach to European procurement decisions appears arbitrary and gives the impression of discrimination against foreign suppliers. From the European perspective, however, there exist strong reservations about advocating explicit market-opening measures, including the abandonment of offsets, without guarantees of reciprocal, improved access to the large U.S. domestic market. European access to the U.S. defense market at both the system and subsystem levels is currently obstructed by a number of legislative obstacles, including defense mobilization restrictions and the inflexibility of U.S. "Buy American" legislation. In the United States, therefore, newly merged U.S. companies operate from a secure and large domestic market and could come to dominate the fragmented European market. They already enjoy a healthy surplus in transatlantic defense trade (see figure 5). In fact, the status quo can still make strong economic sense from the U.S. perspective. Why should U.S. companies have to accept a rationalization of U.S. procurement legislation that might weaken their competitive position in the domestic U.S. market when they currently enjoy good, if imperfect, access to fragmented European defense markets?

Figure 5: Transatlantic Defense Trade

Growing disparities in transatlantic defense trade engender a perception in Europe of an uneven playing field, which threatens to undermine existing U.S.-European defense cooperation and to polarize defense markets on both sides of the Atlantic.

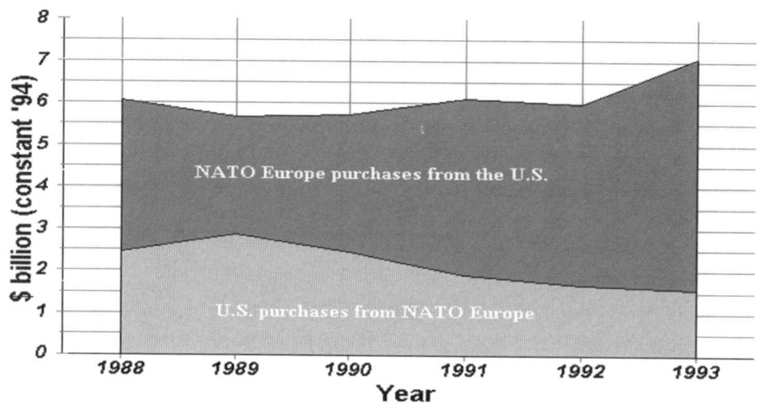

Sources: U.S. Department of Defense. Office of the Under Secretary of Defense for Acquisition and Technology. "FY [1988 through 1993] MOU Defense Procurement Activity Summary," Washington: OUSD(A&T)DP, 1989–1994.

Until there is a more balanced relationship between the United States and Europe in terms of market integration and industrial consolidation, there will be little legislative or corporate support on either side of the Atlantic for far-reaching positive amendments to existing U.S. and European procurement policies and practices.

CHAPTER 2

Recommendations

Clearly, there are many structural obstacles to transatlantic defense cooperation. If U.S. and European governments and companies are to reap the benefits of cooperation, they must develop strategies that take into account the extent and depth of these structural obstacles. For this reason, the Atlantic Partnership Steering Committee advocates a twin-track approach. U.S. and European governments should separate the immediate steps that can be taken to improve the market environment for transatlantic collaborative programs (those that involve cooperation from the development phase onward) from the longer term policies that can create a more positive climate for all other aspects of defense cooperation, including defense trade.

On the first track, the Steering Committee proposes the immediate creation of a Framework Agreement for transatlantic cooperation that would insert specific incentives for governments and companies to see worthy collaborative programs through to fruition. On the second track, the Steering Committee advocates a policy of constructive engagement among U.S. and European governments that will encourage a transition to more competitive, coherent, and transparent defense markets on both sides of the Atlantic.

A New Framework Agreement for Transatlantic Collaborative Programs

U.S. and European governments have reaffirmed their interest in pursuing certain defense programs on a collaborative basis, that is, from the development phase onward. If such programs are to succeed where others have failed, they must provide industry with incentives to offer the best value solutions in terms of technology development, production, and program management. To this end, our central recommendation is that U.S. and European governments should immediately develop, and their legislatures approve, a Framework Agreement for collaborative programs. The Framework Agreement would create incentives by overcoming several of the key obstacles to collaboration.

Although governments would negotiate the exact parameters for national participation in collaborative programs on a case-by-case basis as they do today, pursuing the programs under the suggested Framework Agreement would provide a prearranged mechanism for overcoming or bypassing many of the structural obstacles associated with collaboration.

Prerequisites to Proceed Under the Framework Agreement

A collaborative program would first have to satisfy several prerequisites to be eligible to proceed under the Framework Agreement. Participating governments must be able to

- identify a potential program for collaboration. To conclude that the program would be suitable to proceed under the Framework Agreement, governments would have to be satisfied that it
  - meets a carefully coordinated set of common military requirements;
  - is based upon technological complementarity and avoids duplicative R&D, production, and other costs; and
  - improves the standardization and interoperability of the forces of participating governments.
- negotiate the rough parameters of national investment and returns, as well as the selection process for the program. Governments must generally balance the cost and efficiency benefits that flow from competition in defense procurements against the impact that such decisions have on the national defense industrial base. Given the different national approaches to achieving trade-offs between these two factors, it may be necessary to accept a "managed" approach to competition in the medium term, enabling participating countries to receive a return on the program in rough proportion to the national investment.
- agree on the arrangements for exporting the final system. This would involve
  - establishing a common set of destination countries to which industry would receive, in principle, approval to export; and
  - accepting that exports to any additional countries would have to receive the unanimous approval of participating governments. Under these arrangements, participating governments should commit only to withhold their approval of exports in cases in which vital national security interests are under threat.
- secure approval from their legislatures to proceed with the collaborative program under the terms of the Framework Agreement. The approval process would involve the submission of the rationale for collaborating on the program, including the export arrangements. If national parliaments are satisfied, they would authorize their executives to proceed with the program under the terms of the predetermined Framework Agreement.

Terms of the Framework Agreement

The central tenets of the Framework Agreement should be as follows:

- Governments coordinate as much as possible their disparate technology transfer and security rules and procedures, so that U.S. and European companies can share technology under normal industry contractual and proprietary rules. A general lack of agreement on security laws and rules are a front-end disincentive to multilateral collaboration. The development process, especially in the concept formulation and predevelopment stages, can take potential partner companies beyond the technology for which they have a license. Outside clearly structured, specifically "fenced" areas—such as nuclear and cryptographic technologies or electronic counter-countermeasures (ECCM)—companies should not need to obtain an export license just to discuss a program with a prospective alliance partner. Companies will not give away technologies that are vital to their future competitiveness.

  The simplification of export control procedures within the alliance could be undertaken under the following preconditions:

  - Allied companies would first apply to their respective governments for qualification to receive, in principle, prior approval for export within the alliance.
  - Allied companies would notify their respective governments after the fact when a transfer has occurred to another NATO company or country.
  - Technologies transferred freely within the alliance could be transferred to an outside destination, with the prior approval of the originating country and a proper export license. Companies that abuse this system would lose their license-free status.

- Participating companies take the lead in determining the distribution of work content, technology development, and program management, including arrangements for prime contractorship—in accordance with the parameters for national investment and return that governments have negotiated up front. Negotiations between governments on the parameters of national investment and return on the program should interfere as little as possible with the ability of U.S. and European industry to determine the allocation of work among participating companies so that they can offer the best collaborative arrangement for the requirement. In the United States, allowing European companies to bid as primes on requests for proposals (RFPs) is consistent with the Department of Defense's (DOD) stated objective of purchasing defense systems from the best-value supplier.

- Governments agree to take measures to assure stability and continuity for the band of programs through clearly stated funding milestones, including the financial liabilities for government that end their participation outside the predesignated milestones. Multiyear (Europe) versus annual (United States) approaches to funding insert fundamental instabilities into collaborative programs. Strict commitments to funding milestones can overcome the impossible task of harmonizing U.S. and European legislative approaches to defense programs oversight.
- Governments agree to an accounting methodology for national contributions to the program, including an assessment of the impact of currency adjustments.
- National legislatures allow participating companies to export the collaborative system without further legislative restrictions to the third country destinations that the executive arms of government have identified and agreed upon in advance. If the executive branches of governments can agree on an export regime for a collaborative program and can secure legislative approval for that regime, national legislatures should waive the capacity to impose post hoc restrictions on exports.

Supporting the Framework Approach

For governments to get the best return from the Framework Agreement approach, they should take two further steps:

- Governments should enhance the level of visibility in the requirements-setting process. The government-to-government negotiations on requirements prior to the development of a cooperative Memorandum of Understanding (MOU) must be as transparent as possible if industry is to respond effectively to the growing lead times associated with new technological demands. Fair and timely industry access could involve, where appropriate, legal notification requesting industry's input into the process of drafting cooperative MOUs and classified briefings for industry by NATO and/or national ministries of defense.
- Governments should create a multinational training program for program managers who will oversee U.S.-European collaborative efforts. U.S. and European governments should better prepare program managers for the unique challenges of multinational programs. The current curriculum offered by the U.S. Defense Systems Management College, for example, does not sufficiently train program managers in the complexities of collaborative programs. As collaborative programs become more comprehensive and complex, moreover, the need for training will grow.

Putting the Framework Agreement Into Practice

Establishing the Framework Agreement will require considerable preparatory work. First, U.S. and European defense ministry representatives must come to agreement on the broad contents of the Framework Agreement. They then have to consult with national legislatures to secure the necessary political support for the approach. If such support is forthcoming, national legislatures must pass the necessary enabling legislation to permit the waiver of specific pieces of existing legislation for collaborative programs that are to proceed under the Framework Agreement. Finally, the executive branches of U.S. and European governments must finalize and sign a matching set of bilateral agreements. This will require an important amount of coordination not only across the Atlantic, but also among European governments.

A Policy of Constructive Engagement between the U.S. and European Defense Markets

As discussed above, current imbalances between the U.S. and European defense markets will undermine any far-reaching initiatives to improve the market environment for all forms of defense cooperation outside collaboration. The Steering Committee believes, however, that we cannot sit back and simply wait for this balance to emerge. There are three reasons to accelerate the process.

First, collaborative programs will be few and far between. Given the restrictions on defense budgets on both sides of the Atlantic, the main thrust of defense modernization and procurement in the coming decade will be upgrades, not the development and production of complete new platforms or systems. In this context, governments on both sides of the Atlantic would benefit from the ability to procure the most cost-effective and technologically suitable components and subsystems to sustain the combat efficiency of their military forces. In many cases, U.S. or European solutions will be the right ones for each other's requirements. Just as important, the bonds that exist between Europe and the United States mean that U.S. and European companies could fulfil such requirements while developing relationships as trusted suppliers to each other's defense markets.

Second, governments and industry in the EU are moving from a culture of fragmentation in the defense industrial sector to one of greater integration. This transition offers some important opportunities for transatlantic defense cooperation.

- More effective European coordination of national military requirements will enable European governments to act as effective interlocutors with the United States on collaborative programs.
- Through continued consolidation, European companies often will be in a better position to act as genuine partners with U.S. companies on cooperative programs.
- Greater consistency among national European procurement practices and

policies will allow both European and U.S. companies to operate in a more transparent and predictable market environment.

- By coordinating and formalizing their disparate domestic source preferences and procurement policies at a European level, European governments and the United States may reach the point at which they can negotiate reciprocal market-opening measures.

U.S. government and industry, therefore, have positive reasons to want to be constructively engaged in discussing with their allies the impact of European defense industrial integration on transatlantic defense cooperation.

Third, constructive engagement will also help lessen the risks that are inherent in this transition to a more coherent European defense identity. Existing budgetary pressures and patterns of industrial consolidation mean that the United States and Europe are under pressure to develop and procure defense systems under conditions of confrontation and exclusion rather than cooperation and competition. The growth of interlocking cross-border European industrial interests, combined with deep-rooted perceptions in Europe of an uneven playing field in transatlantic defense trade, means that existing patterns of U.S.-European defense cooperation based on large sales of U.S. defense equipment to NATO Europe and on U.S.-led coproduction deals are not sustainable in the long term. Without an effective U.S.-European dialogue on defense cooperation, therefore, U.S. defense manufacturers may see their traditionally high degree of access to European defense markets diminish as European governments come under increasing pressure to strengthen their domestic defense industries to be competitive with U.S. firms.

To avoid the risks of a new polarization between U.S. and European defense markets, governments on both sides of the Atlantic could undertake some useful steps.

In Europe

As European governments develop a more integrated defense identity, they should ensure that European defense markets remain open to competition and cooperation with the United States. To this end, European governments should

- avoid instituting any new forms of domestic protectionism.
- cooperate to develop a more coherent, integrated, and transparent European defense market, as is currently being discussed in the WEU, providing that any new common external European barriers do not constitute an overall increase in the obstacles facing U.S. defense imports.
- continue the process of rationalizing and privatizing European defense companies. State ownership of European defense firms undermines the case for a further opening of the U.S. defense market in the eyes of U.S. legislators. It also reinforces the perception U.S. companies have that they often operate in a "rigged" European defense market.

In the United States

The U.S. government should make clear its interest in a competitive defense market that is more open to the companies of allied governments. To this end it should

- reject attempts to introduce new legislation that would insist on sole-sourcing from U.S. companies in specific sectors of the defense market.
- revise U.S. "critical items" lists and other mobilization base restrictions to allow European systems to compete in a broader range of U.S. program requirements, especially in those areas in which European companies offer technological advantages and in which the United States does not have the technological capabilities or assets. These restrictions have guaranteed large portions of defense procurements to U.S. industries only. Under Secretary of Defense for Acquisition Paul Kaminski is revising the U.S. critical items lists to apply to a smaller portion of defense procurements, and the DOD leadership is considering further liberalization. The Task Force supports these efforts.
- strengthen and extend the exemptions from "Buy America" legislation for allied countries. The U.S. Buy America Act includes an exemption for European nations that conclude a memorandum of understanding for collaboration.[7] European governments and companies still consider the statute and other defense federal acquisition regulation supplements (DFARS) to be anticompetitive, however. Fortunately, many of these DFARS, though statutorily based, can be eliminated.[8] The DOD, under Paul Kaminski, is already eliminating a broad range of DFARS in order to simplify the overall acquisition process. The DOD should assess how further liberalization will improve transatlantic defense cooperation.

Collectively

The existence of formal and informal offset requirements for U.S. companies bidding on European programs is tied to U.S. restrictions on the ability of European companies to bid for U.S. programs as prime contractors. Only a reciprocal effort by governments and industry on both sides of the Atlantic will ease these obstacles to the emergence of a more efficient and competitive alliance defense market.

---

7. See 41 U.S.C. 10b-1(d).

8. The Buy America Certificate requirement (DFARS 52.225-01), the Buy American Act Supplies requirement (DFARS 52.225-03), and the Buy America Act-Trade Agreements Act-Balance of Payments Program requirement (DFARS 52.225-09) are a few obvious DFARS from which European primes should be exempted.

CHAPTER 3

Conclusions

There is an underlying message in this report. If, at a minimum, U.S. and European governments and industry are unable to launch a small number of successful transatlantic collaborative programs over the next decade, then the chances of doing so in the longer term are negligible. Under present market and regulatory conditions, European and U.S. defense consolidation is likely to proceed along separate and, ultimately, adversarial tracks. U.S. and European industry will shy away from offering joint solutions to the individual and collective needs of allied governments unless the political and legislative environment is one that creates incentives for the success of collaborative programs. At the same time, budgetary pressures for improved levels of U.S.-European cooperation in defense trade and coproduction may prove an insufficient impetus for progress.

Without the establishment of a more constructive relationship among the United States and Western European governments in the defense cooperation field, the Atlantic Alliance as a whole must accept wasteful and unnecessary duplication of R&D and production. The U.S. and European governments must also accept adversarial defense markets that are likely to be incapable of providing our armed forces with the full range of defense systems needed to meet future threats. How can allied governments and industry avoid these perils and achieve a new level of trust?

The speed of further European defense integration—at the military, policy, and industrial levels—is uncertain, but it will not proceed quickly. At heart of the Steering Committee's conclusion is the belief that it will be impossible to develop far-reaching solutions on reciprocal access for U.S. and European companies to each other's markets until this process is further advanced. During this period, therefore, we must make the most of available opportunities, avoid raising false hopes, and minimize any sources of acrimony. Governments and legislators on both sides of the Atlantic need to concentrate on developing a twin track for defense cooperation.

On the one hand, they must make a concerted effort to ensure that a more coherent European defense identity emerges in such a way as to offer the best chance possible for a more balanced transatlantic defense market environment to emerge. This means resisting protectionist forces on both sides of the Atlantic and encouraging a rationalization and greater standardization of procurement policies and practices. On the other hand, U.S. and European governments must also establish some form of Framework Agreement that will allow industry to offer collaborative solutions to existing and future defense needs. Without building a more constructive framework for transatlantic collaborative programs in the near term, cooperation in the longer term among U.S. and European defense companies will become increasingly difficult.

CSIS provides the following annexes to complement the work of the Atlantic Partnership Steering Committee. These analyses and information are the work of CSIS researchers alone. It should not be inferred that any member of the Atlantic Partnership Steering Committee agrees with or supports the ideas contained in these two annexes.

ANNEX A

Program Areas for Collaboration

Daniel Gouré

In 1995, as part of the Atlantic Partnership project, CSIS conducted a series of workshops to assess the possibilities for transatlantic collaboration in specific program areas. This was not an effort to second-guess decisions that governments have already made, for example in the medium extended air defense system (MEADS) missile defense program. Rather, CSIS intended to take a long-term perspective on collaboration and set out to ask what are the opportunities for future defense collaboration. The areas CSIS selected for exploration were land combat systems, battle space management, logistics and lift, next-generation combat aircraft, and theater missile defenses. CSIS intended this effort to be exploratory and to answer three basic questions:

- What kind of requirements might emerge over the next decade that will open up new opportunities for collaboration?
- What kinds of unique or specialized capabilities might each side of the Atlantic bring to the table?
- What kinds of activities should be initiated in the near-term to develop both the technical basis for long-term collaboration and the habits of cooperation?

CSIS did not ask the participants of the working groups to define future military requirements; that is the province of national governments. CSIS challenged them, rather, to think about the evolution of technology, threats, and operating environments in these five program areas to suggest future demand for new capabilities. In some cases, the demand for new systems already exists in long-range acquisition plans and opportunities for collaboration are apparent. In others, new challenges and new technologies have opened up the possibility for major advances in military capabilities while concurrently challenging our thinking about how U.S. and European governments and industry can and should collaborate in the long term.

CSIS instructed the members of the working groups to address the kinds of expertise and skills that their respective national defense industrial bases and governments might contribute to a collaborative program. Both the nations of Europe and the United States possess a wealth of scientific, technical, engineering, and manufacturing skills that can be applied to new collaborative programs. In a number of critical areas, national industrial and technology bases possess unique capabilities that, if married with complementary capabilities in other countries, might produce major technological advances in and costs savings for defense programs.

Because this program-area inquiry sought to look beyond existing systems to what might follow, CSIS recognized from the start that the definition of collaborative activities must start with research and development and not be limited solely to "cooperative" activities like coproduction, reciprocal purchasing arrangements, and foreign military sales. The practical form that cooperation will take is likely to vary from program to program. There are a variety of new collaborative arrangements currently under way, including those for the current definition phase of the MEADS program and the U.S.-led Joint Advanced Strike Technology (JAST) program. Because R&D resources are becoming extremely scarce in many countries, yet the demands for new capabilities continue to increase, engaging in collaborative program development from the requirements stage onward offers an important means by which U.S. and European defense planners can meet future military needs.

Land Combat Systems

The land combat systems working group focused on the family of combat vehicles, including armored fighting vehicles (AFVs), scout and reconnaissance vehicles, self-propelled artillery, and combat support systems such as engineering equipment and logistics support vehicles.

Promising Areas for Cooperation

- ❒ Main battle tank technology demonstrator
- ❒ Armored fighting vehicle self-defense systems
- ❒ Electronic connectivity
- ❒ Identification Friend or Foe (IFF) systems
- ❒ Air ground surveillance
- ❒ Robotics

The principal focus of this program area has always been the main battle tank (MBT). Such systems have been central to modern warfare for more than 50 years. MBT development remains, in the perception of many people, one of the crown jewels of national defense industrial capability, symbolizing not only a method of modern warfare but also of the technological and industrial strengths that represent national well-being in the industrial age. The particular reverence in which governments hold the MBT makes it more difficult than in other cases to establish a program for the collaborative development and joint production of an MBT.

Because several governments have independently pursued national MBT programs, there are considerable discontinuities in national MBT acquisition cycles, which makes it difficult to schedule a future collaborative program. The United States completed acquisition of the M1A1 in April 1993; since then it has undertaken an upgrade program to change a number of A1s to the A2 configuration. The French government is outfitting its armored units with the new Leclerc MBT, and the United Kingdom recently has committed itself to acquiring the Challenger II MBT. The United States, the United Kingdom, and France are unlikely to need a replacement for some 20 years. In addition, the force structure reductions undertaken by the Atlantic governments has reduced the need for new MBTs as older equipment is removed from the inventory. The average age of MBTs in national arsenals has been steadily declining over the past five or six years, further delaying when replacement systems will be required. Only the Federal Republic of Germany (and other European governments that acquired the German-developed Leopard II tank in the 1980s) are likely to begin development of a follow-on system by the middle of the next decade.

Although political concerns and acquisition cycles make a collaborative MBT program unlikely, there are compelling reasons to consider collaboration in research and development on key technologies and components for future MBTs. The development and production of a next-generation armored fighting vehicle—and MBTs in particular—face severe funding obstacles in the current and projected budget environment. As the character of AFVs has changed over the past several decades—with improvements in fire control systems, sensors and command, control, and communications (C^3) capabilities, engines, and armor—the development and acquisition costs of AFVs and MBTs have risen sharply. The improvements foreseen for existing land combat platforms, involving increased employment of more complex and capable electronic systems, will further increase costs. The cost escalation is due, in large part, to the digitization of the battlefield. Although the microchip and the flat panel display have wrought a revolution in avionics; navigation; command, control, communications, and computers (C^4I); and weapons systems, they have also led to an increase in the cost of the overall weapons system. Historically, such technical collaboration has often proved extremely fruitful.

Furthermore, new operational concepts and changes in the threat are likely the generate requirements for next-generation AFVs that may challenge both technologists and budgeters. AFVs have developed increased lethality and survivability on the battlefield over the past several decades, but so too has the threat. Anticipated improvements in the threat in the years to come—deadlier and longer range precision-guided munitions, advanced antiarmor helicopters, an array of brilliant submunitions, and advanced antivehicle mines—raise questions regarding the survivability and effectiveness of AFVs in general and MBTs in particular on the battlefield of the twenty-first century.

Anticipated changes in operations also suggest the need to rethink design concepts for AFVs. In the past, the central problem U.S. and European engineers faced in the design of AFVs was how to fight and win when on the defensive and outnumbered. This put an enormous premium on defensive measures, on armor, and on gun range; mobility was an important but secondary consideration, particularly for MBTs. Although heavy tanks made sense in the context of NATO defense, they raise some serious problems for power projection operations. It is extremely difficult, for example, to deploy a significant MBT force by air. Prepositioned equipment and sea lift are the only viable options. By contrast, a lighter yet survivable MBT would greatly simplify the projection problem for the United States and the out-of-area deployment problem for several European governments.

One of the critical issues facing the land combat systems community is the future of the tank. The "innards" of the combat vehicle—the black boxes that allow tank crews to see, navigate, communicate, and fight—already account for the majority of the development costs. National requirements for the next-generation tank require a revolution in several technologies and, hence, the prospect of a major increase in unit price. Designers can envision a future tank that has both passive sensors and active means to detect and defend against portable and airborne threats. The requirement to defend against advanced antitank helicopters and all manner of smart munitions inevitably raises the price and complexity of the system. Not only are defensive improvements possible, but liquid propellants and electromagnetic gun systems hold forth the promise of entirely new offensive armaments suites. So too does the development of top-attack antiarmor and antiaircraft missile systems for deployment on MBTs and other AFVs. But new offensive capabilities will also be costly. Will nations accept a tank with a $10 million price tag that it essentially invulnerable to current antitank weapons, regardless of their point of origin?

By comparison, other areas covered under the land systems program are potential candidates for a fully collaborative program. Another relatively low-cost area for transatlantic collaboration is in the interoperability of C^4I systems. "Electronic connectivity" will be critical to coalition operations on the digitized battlefield. The transatlantic community should make a major effort to establish common standards for interfaces and the protocols for all combat information systems, from navigation and communications to targeting and kill assessment. Although such an effort may not substantially reduce the costs of the components themselves, it will greatly facilitate the ability of coalition members to operate and fight together.

The most obvious and least controversial area for collaboration is that of combat identification, or Identification Friend or Foe (IFF). Lack of a reliable, common IFF capability led to some tragic friendly fire incidents during the Persian Gulf war. As coalition operations become more common, involving the presence on the battlefield of a wide variety of platforms with adversary and friendly forces alike deploying the same types of combat vehicle, IFF will be a critical element of an international battle management system. There are many national programs underway and the Conference of National Armaments Directors (CNAD) is working on this issue as well. But the issue is so important, both for coalition operations and for national activities in hostile terrain, that governments must assure that, at a minimum, national IFF systems use standardized broadcast protocols. One useful step would be to include European governments in the DOD's combat identification advanced concept technology demonstrator (ACTD) that is testing air-to-ground, ground-to-ground, and noncooperative target recognition combat identification systems.

The future of land combat systems is intimately tied to the development of battlefield C^3I capabilities. Ground surveillance is a related technology that could enhance the effectiveness of current and future land combat systems. Current efforts to develop a NATO airborne ground surveillance (AGS) capability constitutes one part of such a support program (see section on battle space management). The other is real-time data transmission to ground forces. A robust AGS capability, if built around one or more of the current sets of proposed systems, could provide significantly enhanced theaterwide and tactical intelligence.

In light of current and anticipated international activities by NATO, countermine technologies may be a very fruitful area for collaborative work. Mine countermeasures are underfunded and could benefit from collaboration. Operation Joint Endeavor in Bosnia, involving NATO and other governments, demonstrates both the problem posed by mines to peacekeeping forces and the lack of adequate countermine systems. There are a wide variety of areas to be explored including vehicle-mounted, portable, rotary-wing-deployable, and remotely operated systems. Efforts in this direction could also support the activities of the United Nations and other international organizations dealing with the global problem of antipersonnel mines.

Another area identified as a possible focus of collaborative work is that of advanced scout and nuclear, biological, and chemical (NBC) weapon detection vehicles. Survivable, combat-capable scout vehicles are important to armored warfare. So too are reliable NBC weapon detection systems. During the Persian Gulf war, the coalition had to acquire Czech NBC weapon detection systems to supplement the small number available to the coalition members. Several European countries have done extensive research in this area and have deployed systems. Additional efforts, most likely under European leadership, would address current shortfalls in NBC weapon detection capabilities on both sides of the Atlantic.

Finally, military robotics is an important and growing area in which collaboration may prove fruitful. The research on these technologies is still in its infancy, but robotics have potential applications across a range of land combat systems. There is a close linkage, for instance, between the need for realistic trainers for future land combat system operators and the design of partially or fully automated, unmanned vehicles. Likewise, robotic vehicles may provide cheaper and safer countermine capabilities. A modular design approach to robotic vehicles might be an interesting collaborative project. Such a project could explore future designs of combat vehicles to allow a system to operate with humans, with a reduced human presence, and without humans. It was noted, however, that the C^4 requirements to control an unmanned vehicle properly are usually greater than if a person sat in the cockpit or turret.

Battle Space Management Systems

Battle space management (BSM) is the process of gathering information on the battlefield, identifying targets, and allocating battle resources. It includes C^3, battle management (BM), and an array of sensor capabilities. CSIS chose to exclude from its definition the mission of "information dominance," or the denial of information to the enemy—battle space management's corollary in total information warfare and battle space dominance. This definition provided the working group with a finite set of operational requirements and technologies to identify promising areas of U.S.-European cooperation.[9] The term "battle space management" itself is a product of revolutionary changes on the modern battlefield. As shown during the Gulf war, modern armed forces equipped with precision-strike munitions and near real-time surveillance, threat assessment, and command and control capabilities can amass greater firepower at critical junctures of operations than ever before.[10]

Promising Areas for Cooperation

- ❒ Common standards, interfaces, and protocols
- ❒ Airborne ground surveillance
- ❒ Space-based surveillance network
- ❒ Artificial intelligence

But this area encompasses a breadth of activities that pose some problems with technology, development, and cost. It should be recognized that many potential collaborative battle space management programs cross boundaries with the other program areas CSIS has delineated. For example, combat IFF is critical both to battle management and to land combat, and missile defense networks will require sophisticated battle space management networks. Indeed, the uncertain boundaries of the subject of battle space management systems is one indication of the extent to which information systems will dominate all aspects of future warfare. Beyond this recognition should be an appreciation that how system designers, military services, and defense planners think about information dominance will go a long way toward determining how battle space management systems and other information capabilities are developed and incorporated into military forces.

Looking across the Atlantic, it is quickly evident that battle space management is one of the most fragmented areas. NATO alone has at least five major battle space management organizations. In addition, there are national and even service-based architectures and programs that further complicate the development of coherent unitary battle space management systems. The desire to expand coalitions to include non-NATO partners—as is happening in Bosnia—adds to the babble of incomprehensible and incompatible information systems. This raises the reasonable concern among defense planners and policymakers alike that coalition forces may be less able in the future to operate together if they become more dependent on incompatible approaches to BSM.

The critical issue for BSM is the interoperability of multinational information

9. Although they are a critical part of battle space dominance, CSIS chose to exclude national strategic intelligence assets, delivery systems, precision-strike munitions, and weapons sensors from its deliberations.

10. See Michael J. Mazarr, Jeffrey Shaffer, and Benjamin Ederington. *The Military Technical Revolution: A Structural Framework: Final Report of the CSIS Study Group on the MTR* (Washington, D.C.: Center for Strategic and International Studies, 1993), 55–63.

sources and the seamless transmission and use of information from all sources. The collection of tactical and operational information on the battlefield can continue to be conducted by nationally owned sensors. For multinational forces to utilize that information, however, the allies must collaborate to develop common "fusion centers"—nexuses of multidomain smart sensors, secure wide-band communications nodes, and artificial intelligence systems for multisensor data fusion. An integrated fusion center offers the possibility of real-time, nearly autonomous target detection, recognition, acquisition, and even fire control.

To create open and even transparent information architectures for multinational forces, U.S. and European governments must create standards, protocols, and interfaces to allow nations to pass data rapidly between national systems with the confidence that it will be received and understood at the other end. It matters little what symbol is used; rather, it is critical that any choice of symbol on the map convey the same message about the situation to all parties. To this end, governments need to pay attention to collaboration in the technology and software for data fusion. An easy first step is to create a transatlantic committee on standards and protocols that should seek to deconflict the many different approaches currently being pursued and that would ensure—without affecting the ability of individual firms or governments to design or acquire preferred systems—that different systems operated by different nations could be integrated into a single coalition battle space management architecture.

One of the most critical issues in the future of battle space management is airborne ground surveillance. A concept of operations for battle space management will govern governments' choices for an architecture of surveillance, intelligence processing and distribution, combat identification, operational planning, attack warning, and combat analysis systems. It is a prerequisite to harmonized systems requirements and rational, cost-effective procurement of next-generation systems. For this reason, the members of the alliance must decide whether to deploy an entirely new NATO airborne ground surveillance (NAGS) system or to integrate existing systems, including the U.S. Joint Surveillance and target attack radar system (JSTARS), France's Horizon and Italy's Creso heliborne systems, and the United Kingdom's airborne stand-off radar (ASTOR) platform. NATO will certainly acquire at least one of the these systems and may exploit their complementary aspects to acquire a mixed NAGS capability. Governments must make further efforts not only to make NAGS a reality but also to exploit further its potential to support the expanded range of missions being conducted around the world by U.S. and European armed forces.

A key problem for NATO operations, beyond adequate and open communications, is deconflicting the battlefield. This is necessary to prevent shooting at friendly forces or interfering with engagement of hostile forces. The Persian Gulf war demonstrated the appalling inability of coalition forces to communicate with each other and to distinguish friendly from enemy forces in the confusion of the battlefield. The United States, for example, suffered from the different services' incompatible communications and IFF systems. Differences in secure voice air-to-air communications impeded AWACS control of navy fighters, and the daily Air Tasking Order had to be physically couriered to the navy's aircraft carriers because it could be transmitted only through air force channels. Multinational operations will likely produce exponentially more communications difficulties like these unless U.S. and European governments rationalize standards and protocols, create a harmonized concept of operations, and establish a multinational airborne ground surveillance architecture.

Another promising area for collaborative work is shared space-based intelligence. Space-based systems provide situational awareness to all forces. Increasingly, more nations have independent access to such intelligence, though only a few nations maintain national surveillance programs. The development of an independent European space-based surveillance capability is a noteworthy change in the environment; this technological base in Europe complements the capabilities of the United States. A critical requirement for the next century will be to further the development of space-based surveillance and to ensure that the results of such activities can be shared among the future coalition partners. A coordinated effort to ensure transmission and receipt of data would be a useful first step in the creation of a multinational space-based surveillance network.

Artificial intelligence (AI) is an important factor in the exploitation of the information revolution. Artificial intelligence may be the only means to manage the mass of data that will be transmitted from the wealth of sensors available to coalition forces in the future and to ensure that the right data gets to the correct recipient at the proper moment. There is currently a lot of work under way in this area. It would be desirable to avoid duplication in these efforts as well as to promote the development of common AI programs.

Theater Missile Defense

In its assessment of theater missile defense (TMD) systems, CSIS includes defense against ballistic missiles, cruise missiles, and antiship missiles. This area requires more than collaboration in the development of hardware, although that is not a small part of future transatlantic activities. It also involves cooperation on the resolution of requirements and threat perception issues.

Promising Areas for Cooperation

- Overall TMD architecture
- Boost-phase intercept technologies
- Sea-based TMD systems
- Cruise missile defenses

Some European governments do not see much of a prospect for collaboration in what the United States terms national missile defenses, except possibly in the role of subcontractors. With regard to TMD, there are a number of views in Europe. The German view is that the problem is largely one of extended air defense (EAD). For the Germans—who have the U.S.-developed Patriot antimissile system, are buying the PAC-3 upgrade of Patriot, and are a partner in the multinational MEADS—there is little else that needs to be done in the way of defensive systems, although they would like to collaborate on future technology programs. The French, Italian, and British views were quite different, recognizing that extended air defense would not be sufficient to address the problems posed by even moderately sophisticated ballistic missiles. Moreover, the British and French governments have national military requirements for out-of-area operations for their armed forces and a consequent need for mobile defenses against ballistic missiles.

The MEADS program likely will be the venue for near-term transatlantic collaboration. From the point of view of a number of European governments, the approach to collaboration taken in MEADS is a test case for future collaborative efforts. There are potential stumbling blocks that may prevent the program from actually reaching the production stage, such as conflicting national requirements, funding, progress in key technologies, and difficulties with system integration. Additionally, many in Europe are concerned that the United States is funding secret, or "black," missile defense programs that may eventually torpedo collaborative programs like MEADS. The problem from the European perspective is the inability to convince the "iron majors," or mid-level program officers, in the U.S. procurement system to accept a jointly developed and produced product.

The sharing of intelligence and early warning data may be the most valuable area for transatlantic cooperation. Currently, this would involve sharing U.S. defense satellite program (DSP) data. Many in Europe are concerned about the reliability of the data and guaranteed access. Overall, however, there has been great interest in creating the kind of data links and protocols that would allow a European early warning system to receive, integrate, and use data from U.S. and European sources.

It is impossible to separate theater missile defense from battle management and surveillance. Of critical importance in the creation of a transatlantic defense against missiles is collaboration on the creation of a battle management/C^3 system that allows for the unfettered transmission of data. Collaborative development of interfaces, protocols, and standards to allow separate national systems to access and transmit data may be the centerpiece of the effort to create a transatlantic theater missile defense system. It may be possible to create a program for missile warning, detection, and targeting along the lines of the airborne early warning (AEW) and airborne ground surveillance programs. Indeed, a transatlantic BM/C^3 and early warning capability is a natural complement to the AEW and AGS programs.

NATO needs to address the issue of a TMD architecture. A basis for broad cooperation in TMD could come from the decision to make MEADS the lower-tier element of a TMD system and theater high-altitude air defense (THAAD) the upper-tier element. Europe's commitment to THAAD would be matched by a U.S. commitment to MEADS. Work could then proceed on both systems with the participating nations and firms secure in the knowledge that both systems were likely to be pursued to completion by all parties.

Collaborative possibilities should be explored in the development of boost-phase intercept (BPI) capabilities. Operationally, European governments are interested in BPI because it offers the possibility of intercepting ballistic missiles while minimizing the risk of fallout on European cities. Collaboration in directed energy weapons is only one of several promising areas. The kind of challenges the BPI problem poses would inevitably result in interesting and lucrative breakthroughs in a host of defense and dual-use technologies. Additionally, BPI is a valuable area in which European industry can contribute some critical technologies in engines, sensors, and even materials.

Another area of possible cooperation is sea-based TMD systems. There is interest in the potential European applications of technologies from the U.S. Navy's lower- and upper-tier systems. European governments might be interested in being involved in some of the technology programs, particularly in sensors and hit-to-kill vehicles. The clear limiting factor for making this program a basis for a TMD activity has been the radar. The absence of an Aegis-quality radar meant that the Europeans could not develop a stand-alone capability based on the Horizon frigate. For this system to be the basis for a naval missile defense, either the Europeans would have to acquire Aegis cruisers or the United States would have to provide the fire control for European missiles. The latter case presumes that the BM/C^3 and operational requirements for a multiship, multination, cooperative engagement capability can be developed and is politically acceptable.

One area that has received a great deal of attention in threat assessments but relatively little R&D funding is the problem of combatting sea-launched cruise missiles. Collaborative programs exist in naval defense systems, but for the future, the problem will be one of supersonic missiles (such as the SS-N-22 "Sunburn"). Development of warning systems, high velocity interceptors, and kill mechanisms are all possible areas for collaborative work.

Long-term research in missile defenses should consider the possibility for collaborative work in the development of directed energy weapons. Both sides of the Atlantic bring can bring capability to bear in this area. A collaborative approach might also address the problem of inadequate funding for research on directed energy weapons, which may be the future of missile, air, and anti-cruise missile defenses.

Logistics and Heavy Lift

The adequacy of U.S. and European governments' ability to support forces in the field (that is, logistics) and to move forces to where they are needed (lift) has been a long-term problem on both sides of the Atlantic. Logistics has been a national responsibility, which has often created enormous problems in the coordination, support, and planning for multinational operations. Improving the interoperability of equipment is a way to ease the logistics problem by allowing forces to be supplied from any national logistics system. Admittedly, many cooperative efforts to improve the logistics process have been unsuccessful. These efforts have concentrated largely on the standardization of equipment among U.S. and European armed forces rather than enhancing the flow of men and materials by improving transport and logistics management. It is in these latter areas that the most significant and rapid progress can now be achieved through cooperative endeavors.

Promising Areas for Cooperation

- ❒ Fast sea-lift technologies
- ❒ Water jets
- ❒ Innovative hull designs
- ❒ Multinational logistics systems
- ❒ Fast load-and-unload systems

National governments have largely had responsibility for lift capabilities, with the United States possessing the bulk of long-range heavy lift capabilities. Changing national military requirements in Europe and the United States now demand another approach to these subjects, one based on a recognition of the centrality of out-of-area, ad hoc operations in planning for future military forces. Rapid deployment of forces requires either continual forward presence or intensive capabilities, preferably both. Only the United States has been able to afford the kind of predeployment necessary to provide rapid deployments, and even then only in a limited fashion. The alliance will require improved lift if it is to carry out missions beyond the reach of road and rail transport.

The peacekeeping deployment to Bosnia demonstrates the importance of logistics and lift. Moving heavy forces is a difficult task made even more challenging by the absence of adequate roads, the inadequacies of air and sea heads, and the lack of heavy equipment for loading and unloading equipment. In Bosnia, the availability of overland supply and transport routes, plus the proximity of major intermediate depots to the deployment zone, has eased considerably the pressures on coalition lift capabilities. For future overwater or long-distance deployments, however, improved sea and air lift will be absolutely necessary.

Sea Lift

A revolution is under way in lift and logistics, particularly as it relates to transport of goods and materials by sea. In particular, new hull designs and power systems for ships and new forms of logistics management hold forth the prospect for changing both the deployability and sustainability of forces. A logistics support system based on rapid loading and unloading of ships and high-speed transit could significantly reshape the military strategies of U.S. and European governments. The technological leadership of several European industries in the design and construction of fast sea-lift systems and rapid logistics management makes this possibility for collaboration highly attractive.

U.S. and European armed forces rely upon transport by sea, not air, for the majority of their fighting equipment. This is particularly true for the United States because the U.S. government continues to relocate a number of overseas units back to the continental United States (CONUS). The governments of France and Great Britain both have new military requirements for a global mobility capability for their armed forces.[11] New fast sea-lift technologies may provide the means to redress the current shortfall in lift capabilities. Improvements in sea-lift technologies and practices not only will greatly enhance the mobility of U.S. and European armed forces but will strengthen the credibility of governments' security policy while easing reliance on pre-positioned equipment.

Traditionally, the United States has planned for its heavy forces to be delivered from CONUS to the theater at approximately 20 knots. A number of commercial vessels today, however, take advantage of new hull designs and innovative propulsion systems to improve their speed for delivery of time-sensitive cargoes. A number of current "fast ships" (up to 15,000 tons displacement) with water "jets," or impellers, are capable of more than 45 knots, and designers are planning a transatlantic fast ship of larger displacement. When combined with new hull designs, fast ships with water jets may cut shipping times for the United States to Europe in half.

Fast ships consume a lot of fuel, however, so their operating economics favor only those cargoes that are high in value and urgency. The governments of the United States and Europe should explore what time-urgent military cargoes could be efficiently and effectively transported by fast ship. If U.S. and European governments can identify a prospective joint military requirement for fast ships, then they should collaborate on a fast military transport ship demonstrator. This demonstrator could evaluate the feasibility of marrying water jets, innovative hull designs, and new loading and handling systems. It could also validate the operating economics of fast military sea lift for a variety of bulk, oversized, and outsized cargoes. Fast sea-lift technologies already exist, so fast military transport ships will require shorter development times and less up-front investment. U.S. and European governments also should cooperate to explore the utility of fast ships. If fast ships are used for deployments only in times of crisis or conflict, their cost effectiveness will be considerably reduced.

Air Lift

The United States, under planning scenarios of the Bottom-Up Review, faces a lift shortage of several million ton-miles per day, though the planned procurement of a 120 aircraft fleet of the new C-17 will help.[12] In Europe, a renewed emphasis on out-of-area operations is creating new demands for an indigenous lift capability. Although it is uncertain how many aircraft can be afforded, this in part is the rationale of the Future Large Aircraft (FLA) program.

Although the U.S. and European governments have needs for air lifters, they will find it extremely difficult to coordinate their national military requirements due to a large C-17 program in the United States, strong multinational political commitment in Europe to the FLA, and the fact that these aircraft reflect different

11. See *Statement on the Defence Estimates 1995: Stable Forces in a Strong Britain* (London: Her Majesty's Stationary Office, May 1995) and *Livre Blanc sur la Défense* (Paris: Ministère de la Défense, 1994).

12. John T. Correll, "Anything, Anywhere, Anytime," *Air Force Magazine* 79, no. 2 (February 1996), 3.

priorities for strategic versus tactical lift. Despite these constraints, there is a case that the U.S. and European governments should at least discuss the specifications and requirements for the FLA. Although the United States already has a variety of air lifters, the FLA may offer solutions for a number of future U.S. military needs. First, the FLA may balance out the U.S. air lifter fleet by providing an alternative tactical air lifter; as the U.S. C-130 fleet ages, the United States may face a shortfall of flexible airframes capable of the variety of tactical missions currently executed by the C-130s different versions. Second, the FLA could provide a replacement option for a number of other air missions, such as maritime patrol missions or tanker operations. The United States will need to participate in the requirements and specifications process if the FLA is to play a role in prospective U.S. needs.

Next Generation Combat Aircraft

U.S. and European governments are unlikely to collaborate on entire airframes for the next generation of aircraft. The United States is developing the F-22, an aircraft that will satisfy many of the U.S. Air Force's requirements for the next 30 years. In Europe, cooperation among the United Kingdom, Germany, Spain, and Italy has led to the Eurofighter (currently in development). France is producing the Rafale, and Sweden is producing the Gripen, a smaller airframe than the other two. Because these are new programs, U.S. and European governments are unlikely to begin replacement programs for these aircraft—collaborative or otherwise—for 20 years. The existence of these four programs also reflects the differing national requirements that will make it difficult to develop a common, transatlantic military requirement for a next-generation combat aircraft. Even if governments can agree to a common requirement, it is unlikely that U.S. and European governments and industry will be able to agree upon an acceptable division of technology and production shares. For all of these reasons, collaboration within Europe likely will precede any transatlantic effort.

Rather than cooperate on the development of entire airframes, U.S. and European governments need to collaborate on components and subsystems. Such a collaborative strategy may preserve technological capabilities at a time when the combat aircraft industry rationalizes and consolidates on both sides of the Atlantic. Furthermore, longer procurement schedules and smaller buys may erode technological expertise in companies on both sides of the Atlantic.

Promising Areas for Cooperation

- Thrust vectoring, vertical flight
- Off-boresight targeting
- Sensor fusion
- Manufacturing technologies

- All-weather ground attack systems

Teaming arrangements for components with niche technologies create market pull. For two decades now, British Aerospace's experience with VSTOL aircraft has fostered U.S.-UK collaboration, first on the AV-8B Harrier II program. British Aerospace is now a subtier participant in the McDonnell Douglas team competing for the U.S. Joint Advanced Strike Technology (JAST) program. French companies produce the world's best off-boresight targeting systems for air-to-air combat, a technology for which the DOD has recently issued a requirement.[13] Rockwell International and Daimler-Benz Aerospace have also collaborated on another important niche technology—thrust vectoring—in the X-31 technology demonstrator. Transatlantic complementarities may exist in sensor fusion, low observables, or stealth, IFF, and various manufacturing technologies. These types of teaming arrangements, from subtier participation in national programs to collaborative demonstrators, may prove to be the best way of preserving and enhancing the combat aircraft technology base on both sides of the Atlantic during the increasingly lengthy gaps between development and production of combat aircraft.

One specific area for collaboration is in ground attack systems. The United States has a requirement to replace the A-6 aircraft, while in Europe, Great Britain needs a replacement for the Tornado and France needs one for the Mirage 2000. Although it is unlikely that they will produce a collaborative airframe, U.S. and European governments should take this opportunity to collaborate on components such as weapons systems, navigation systems, and avionics. This way they can produce a highly common aircraft, even if the airframes are different, thus simplifying logistics and maintenance. If governments are to produce affordable airframes, they must not expend scarce monies to duplicate R&D in a number of these specialized technological capabilities. 23

---

13. See Craig Couvalt, "French Expand Attack Options," and Michael A. Dornheim, "U.S. Fighters to Get Helmet Displays After 2000" in *Aviation Week and Space Technology* (October 23, 1995), 45–48.

ANNEX B

The Costs of Collaborative Programs

In general, it is exceedingly difficult to judge the cost effectiveness of multinational collaborative programs. Historical lessons suggest, however, that international collaborative programs cost more than national programs for several reasons. Collaborative programs often have inefficient production arrangements because each partner insists on a share of the production; partners insist upon R&D and production shares, despite having little or no comparative advantage to develop new technical capabilities; unique national requirements for weapons systems or subsystems can mitigate the benefits of economies of scale; and multiple national program offices or oversight organizations can slow decision-making and impede program managers. The historical record shows that these and other factors can contribute significantly to the costs of collaborative programs and that collaborative programs rarely achieve their theoretical economic benefits.[14] Nonetheless, though overall program costs may be higher, a nation's share of a collaborative program often is significantly less than the costs of a national, go-it-alone approach. In today's world of declining defense procurement in Europe and in the United States, a collaborative approach may mean that governments procure systems they could not otherwise afford.

A recent case in which a multinational approach can be less costly than going it alone is the Multifunctional Information Distribution System (MIDS). This tactical data link allows a number of platforms—from helicopters and fixed-winged aircraft to ground vehicles—to share targeting and other tactical information. The United States agreed in 1986 to participate in the MIDS project and signed an MOU with the United Kingdom, Norway, Italy, and Canada in April 1987. Despite these U.S. commitments, the U.S. Air Force continued to develop its own tactical data link, the Joint Tactical Information Distribution System (JTIDS). Although the Air Force undertook an aggressive program to control the unit costs of JTIDS, the current version (the 2R terminal) costs approximately $200,000 more per terminal than the MIDS terminal.[15] The U.S. Department of Defense estimates, furthermore, that as a common NATO system the MIDS program will have lower maintenance and support costs over the life cycle of the program. With plans to procure nearly 1,400 MIDS terminals, the program offers considerable savings on unit and life-cycle costs over the U.S.-only JTIDS program. In August 1995, the

14. See Mark Lorell and Julia Lowell, *Pros and Cons of International Weapons Procurement Collaboration.* (Santa Monica, Calif.: RAND, 1995), 14-17.

15. *Aviation Week and Space Technology*, August 28, 1995, 25.

U.S. Air Force decided to cancel procurement of the JTIDS program in favor of MIDS. By September 1995 the Pentagon agreed to full participation in the MIDS program.

MIDS is an exceptional case; international collaborative programs usually have overall costs that are greater than a national program. But national shares of the costs of collaborative programs often are significantly less than the costs of a national program. The theoretical costs savings arise from the economically efficient allocation of work share. Although the political imperatives of collaboration often undermine this economic benefit—particularly for nations that desire to acquire production know-how or, at a minimum, to spend tax money domestically—collaboration on smaller and component systems or munitions may offer greater opportunities to allocate work more efficiently. Because smaller systems and munitions have higher volume production than weapons platforms, production can be divided with fewer detrimental effects on economies of scale. Furthermore, these programs may be less politically visible or less desirable from the standpoint of national technological capabilities. For these reasons they may be less susceptible to political pressures than would a domestic manufacturing capability.

A number of current programs that are early in their development may offer this type of economy of effort to their partners. The NATO Evolved Sea Sparrow Missile (ESSM), for instance, may offer its partners cost savings because of its great economies of scale when in full production. The Trigat antiarmor missile offers similar cost savings and the advanced short-range air-to-air missile (ASRAAM)/advanced medium-range air-to-air missile (AMRAAM) family of weapons reciprocal purchasing arrangement attempted to take advantage of the efficient allocation of national work shares between two complementary programs. Another interesting example is the U.S. Joint Advanced Strike Technology (JAST) aircraft program. Although JAST may be best described as a national program that derives most of its economies from the rationalization of requirements among the U.S. armed services, the participation of British Aerospace in one of the competing teams suggests that the U.S. government may be seeking a cost- and work-share partner to economize effort once the program enters the R&D phase.

Although overall program costs may be greater, collaborative approaches to many weapons systems may offer U.S. and European governments the only means of procuring these weapons. The French, Italian, and British governments probably cannot afford, for instance, to develop a next-generation frigate on a go-it-alone basis. By collaborating on the development of the Horizon-class frigate, Paris and London not only get a program they otherwise would be unlikely to fund, they also free funds for other modernization and procurement programs. For similar reasons the United States is collaborating with Italy, France, and Germany on the MEADS to provide European governments with a theater missile defense capability. The U.S. European allies benefit from the economies of scale offered by a larger U.S. program, while the U.S. Army (whose proposed Corps SAM [surface-to-air missile] system provides the backbone of the MEADS program) sustains a program that Congress had threatened otherwise to eliminate. Although cost overruns and program delays are common features of collaborative programs, a multinational approach often is the only way U.S. and European governments can afford to meet many of their military requirements. Particularly in today's austere fiscal environment, these costs savings are critical to U.S. and European governments' efforts to meet the broad range of their modernization needs.

All these examples, however, are anecdotal. Economic data on collaborative programs is not readily available, in part because in the past (and in Europe in particular) governments have collaborated for political and military reasons rather than for economic reasons. Even the examples cited here may be the product of somewhat unique circumstances. The MIDS program, for instance, is the product of unusually similar procurement schedules, national military requirements, and high-volume production that offer more opportunity for a rational work-sharing. Furthermore, a number of the programs are in their early phases: the ESSM and Horizon programs are in development, EFA is in testing and evaluation, and MEADS and JAST are in the concept exploration phase. The cost savings of these programs have yet to be realized, particularly those savings offered by rationalized production schemes to which collaborative partners rarely agree. Nonetheless, the anecdotal evidence offers two important and related lessons. First, collaborative programs can offer savings in terms of a nation's share of costs and risks if a program is well managed and partners agree to economic rationalization of production. Second, it is rare that international programs offer the serendipitous concurrence of national military requirements, procurement time lines, and rational work shares. For both these reasons, U.S. and European governments must choose the programs on which they collaborate carefully.

OTHER CSISBOOKS OF RELATED INTEREST

Nordic-Baltic Security: An International Perspective
Arne Olav Brundtland and Don M. Snider, editors
Contributors: Clive Archer, Stephen A. Cambone, Helmut Hubel, Atis Lejins, Sergei Oznobistchev, Trivimi Velliste

In examining the future security arrangements of the Nordic and Baltic areas, the authors look at emerging security needs, especially in terms of possible threats. They outline the security arrangements likely to be available to attempt to deal with these needs and comment on the possible structures preferred by the leaders of the states involved.

CSIS *Report* 132 pp. 1994
ISBN 0-89206-256-8 $16.95

Does France Still Count? The French Role in the New Europe
Steven Philip Kramer

"Lucid and insightful analyses of French foreign and security policies....a major contribution to both academic and public policy...free of debilitating jargon....His monograph deserves to be widely read by American diplomats and policymakers....For those seeking an understanding of post-cold war France, this book is the best place to start."—History of European Ideas

Washington Papers/ 113 pp. 1994
Praeger (Index)
ISBN 0-275-95061-1 (pb) $12.95
0-275-95060-3 (hb) $45.00

Turkey: The Challenge of a New Role
Andrew Mango

"This book is a tour de force and tour d'horizon of modern Turkey by an old Turkish hand. In nine compact chapters the author manages to summarize succinctly the major political, economic, social, and diplomatic developments the country has confronted in the last two decades and the challenges it faces in the final decade of the 20th century. The major focus is on policies to enhance modernizing and developing an export oriented economy."—CHOICE

"Mango offers a number of practical solutions, and his book is a welcome addition to the literature on contemporary Turkey."—Foreign Service Journal

Washington Papers/ 144 pp. 1994
Praeger (Index)
ISBN 0-275-94986-9 (pb) $12.95

OTHER CSISBOOKS OF RELATED INTEREST

Critical Issues in Defense Conversion
A Consensus Report of the Senior Group on Defense Conversion
Harold Brown, project chairman
Daniel Gouré and Debra van Opstal, project directors

The CSIS Committee on Defense Conversion concluded that a comprehensive framework linking defense conversion to the restructuring of the defense industrial base and reallocation of defense investments is urgently needed. The challenges are to meet the nation's basic security needs at a reduced level of spending, to minimize the negative impact on the economy of the build-down, and to leverage the remaining defense investment for national economic and technological goals.

Panel Report 51 pp. 1994
ISBN 0-89206-266-5 $21.00

From Shadows to Substance: An Action Plan for Transatlantic Defense Cooperation
Conference Summary and Background Report of the CSIS Atlantic Partnership Project
The Rt. Honourable Sir Geoffrey Pattie, MP, and Michael T. Smith, Atlantic Partnership cochairs
Daniel Gouré, project director

Based on a conference held on Capitol Hill in February, this report sets the agenda for the future of this important endeavor. The report highlights consensus over mutual U.S.-European interest in the future of defense cooperation with emphasis on the current favorable political climate and the globalization of economic activity. Additionally, the conference identified specific project activities that are conducive to fostering transatlantic defense cooperation.

CSIS *Report* 42 pp. 1995
ISBN 0-89206-276-2 $10.95

Defense in the Late 1990s: Avoiding the Train Wreck
Don M. Snider, Daniel Gouré, and Stephen A. Cambone, project directors

Based on the work of a December 1994 conference in Washington, D.C., this report examines the potential for a "train wreck" in defense: the endangering of future U.S. capabilities by the current overcommitment of U.S. military forces and by the failure to provide a strategy for reshaping the military to meet future requirements. The authors examine the short- and long-term options facing defense planners and propose alternatives for the future.

CSIS *Report* 25 pp. 1995
ISBN 0-89206-316-5 $14.95